STEP BY STEP GUIDE TO

Napkin Folding

Marie-Rose Sorier
Simone Bonroy

SUNBURST BOOKS

PREFACE

Would you like to really spoil your guests?
Astound them not only with your culinary talents but
also by giving your table a truly special occasion feel.
They will admire not only your beautiful china and
your new tablecloth but, above all, your artistically
folded napkins. The secret of your success is this book,
which gives you step-by-step instructions on how to turn
an ordinary napkin into a little work of art.
Whatever the occasion, you will find a pattern here to
suit your table. Some patterns will simply give your
table a more festive look, while others are for more
formal occasions: be it a business dinner, a meal with
friends, a reception, or a childrens' birthday party. A
napkin folded in an original and decorative way never
fails to impress. What is more it often adds to the
atmosphere of the occasion.
With this book and a little patience, you will be
surprised at how skilful you can become.

CONTENTS

INTRODUCTION

*Before presenting the designs
here are some important points
to consider when choosing your napkin.*

A NAPKIN FOR EACH OCCASION?

*Choice is, of course, a matter of personal taste but also depends on the occasion
for which you want to use the napkin.
For special occasion dinners, it is best to cross paper napkins off your shopping
list. Go for cloth napkins. Choose the colour, size, material and type to match
the tablecloth you are planning to use.*

THE IMPORTANCE OF THE FABRIC

*Your choice of napkin will be governed by the occasion and also by the design
you are planning to use. Napkins made of synthetic fabric do not hold their
shape well. So use them for designs where the shape is secured by a napkin ring
or a glass.*

SHOULD I STARCH MY NAPKINS?

*If you have chosen to use cotton or linen napkins starch them well. This will
create the best effect and ensure that the napkins stay in shape.
When you put the napkins away, it is best to lay them completely flat to avoid
fold marks. If you don't have space to do this, starch your napkins just before
you fold them into patterns.*

WHAT SIZE AND SHAPE OF NAPKIN SHOULD I USE?

*The shape and size of the napkin are important too. As regards shape, all the
designs in this book are made using square napkins. Obviously you will not
achieve quite the same result using any other shape.
Paper napkins available in the shops come in a variety of sizes. For the
patterns in this book, especially the more complicated ones, we have opted for
larger napkins, as you will see when you have experimented with some of the
designs.*

PLACING THE NAPKIN ON THE TABLE

Another point to note is the position of the napkin on your table. Here too, you have several alternatives, depending on the occasion and the pattern you choose.

Many people wait until the last minute to lay their table. It might well be worth planning the day before to decide where your napkins will be shown off to best effect.

HOW TO USE THIS BOOK

Finally, a word about the way we describe the designs. The folding of each design is described step by step. Each step consists of explanatory text and an illustration, together with a number of diagrams and symbols, which indicate, for instance, the direction in which you should fold the napkin. You will find the guide to these diagrams and symbols on pages 12–13.

UNDERSTANDING THE INSTRUCTIONS

In order to explain as clearly as possible the procedure for folding the napkins, we have used a number of symbols. When you have studied these instructions you are ready to start work.

1. Broken Line _ _ _ _ _ _ _

   A broken line indicates that the napkin is to be folded forwards in the direction of the arrow.

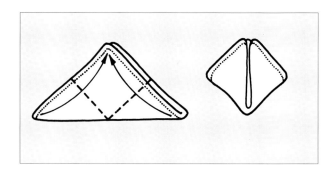

2. Dots and Dashes _._._._

   Alternate dots and dashes indicate that the napkin is to be folded backwards in the direction of the arrow.

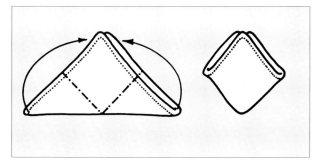

3. Arrow

   An arrow shows you which way to fold the napkin.

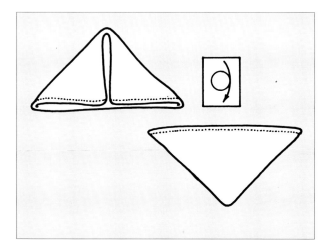

4. Circle ◯

   A circle indicates where you should hold the napkin.

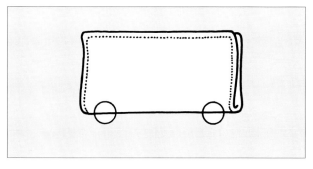

5. Dotted line ....

A dotted line indicates the open edges
of the napkin.

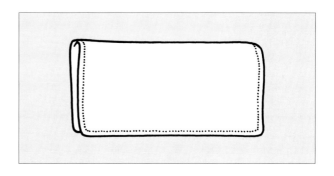

HOW TO FOLD A NAPKIN IN FOUR

In some designs, the instructions begin with a napkin already folded in four.
The following will show you exactly how to do this.

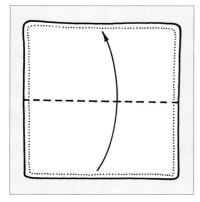

1 Fold the napkin upwards.

2 Fold the napkin to the right.

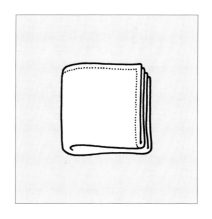

3 The napkin is folded in four. Dotted lines indicate the open edges.

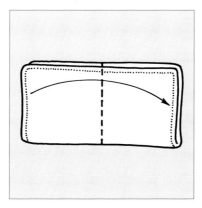

FLAT DESIGNS

*Although less showy, flat designs are every
bit as creative and original as others.
Some designs are particularly pretty.
The advantage of flat designs is that you
can keep your secret until it is time to
serve the meal. Then each of your guests
will discover your miniature masterpieces.
Surprise them with a napkin into which
you have slipped their name card or even
one containing a small bread roll...
Why not make it an oriental evening?
Show off your knowledge of Japanese
culture and present each guest with a
little miniature kimono...
Satisfaction guaranteed!*

This is an elegant napkin to delight your guests!
You could slip a card with their names on inside each one.

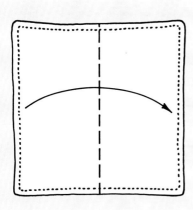

1 Fold the open napkin to the right.

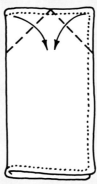

2 Fold the upper corners down towards the middle.

3 Fold the lower part up over the upper so that the point is no longer visible.

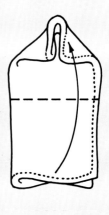

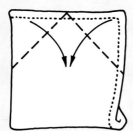

4 Fold the corners of this half down to the middle.

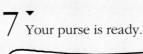

5 Fold the point of the upper layer point downwards.

7 Your purse is ready.

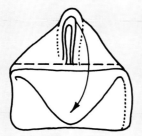

6 Repeat with the other point but leave it a little higher.

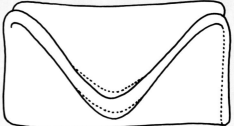

Ideal to accompany a fish dish . . .
This will give your guests a foretaste of the meal.
You will need a lot of patience, but the result is well worth it.

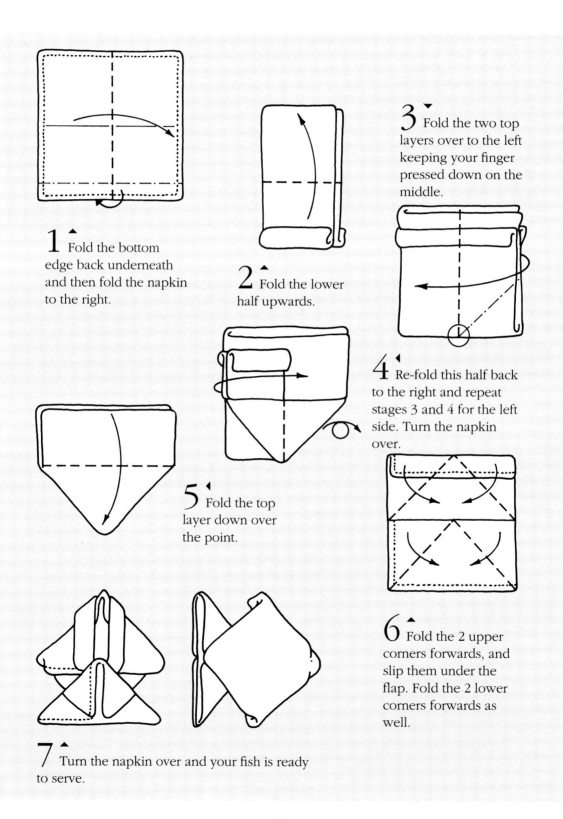

1 Fold the bottom edge back underneath and then fold the napkin to the right.

2 Fold the lower half upwards.

3 Fold the two top layers over to the left keeping your finger pressed down on the middle.

4 Re-fold this half back to the right and repeat stages 3 and 4 for the left side. Turn the napkin over.

5 Fold the top layer down over the point.

6 Fold the 2 upper corners forwards, and slip them under the flap. Fold the 2 lower corners forwards as well.

7 Turn the napkin over and your fish is ready to serve.

A geometric shape that will make your table sparkle...
Mix and match the colours of the napkin and tableware.
A real little gem...

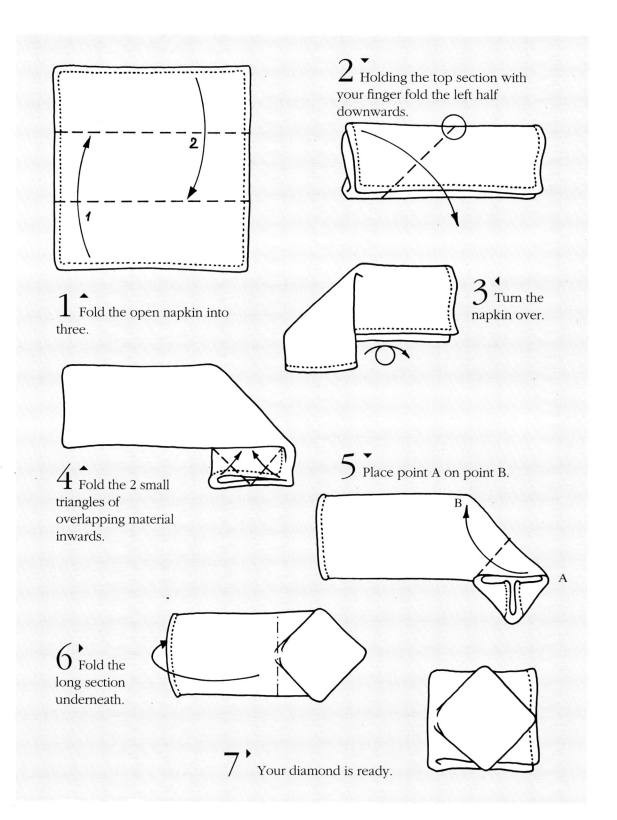

1 Fold the open napkin into three.

2 Holding the top section with your finger fold the left half downwards.

3 Turn the napkin over.

4 Fold the 2 small triangles of overlapping material inwards.

5 Place point A on point B.

6 Fold the long section underneath.

7 Your diamond is ready.

A great idea for an Autumn party…
With the appropriate table decoration,
the effect is stunning.

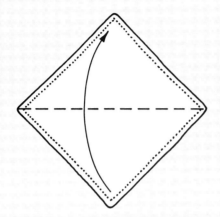

1 Lay out the open napkin with corners at top and bottom and fold the bottom corner upwards.

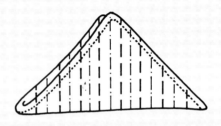

2 Fold the napkin from left to right, concertina-style.

3 In the centre of the concertina, fold one half towards the left and the other towards the right. Turn the napkin over.

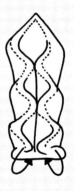

4 Fold a small bottom hem underneath.

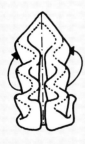

5 Hold together the right and left halves at the back and tuck the edges into the centre fold.

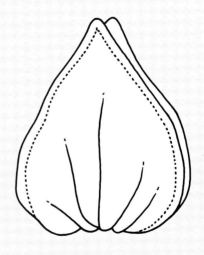

6 Adjust the shape of the leaf.

Planning an important dinner party?
Give your table real style!
Add a little bow tie, the result is perfection.

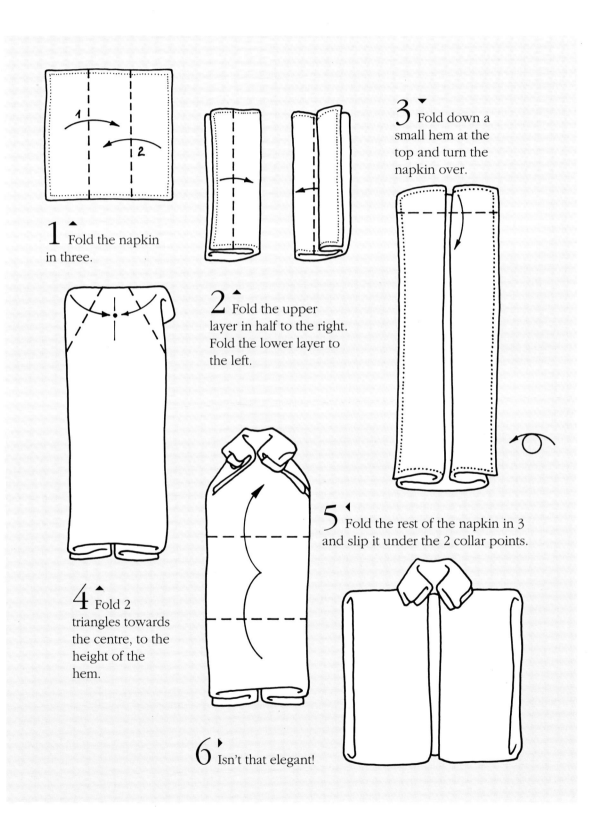

1 Fold the napkin in three.

2 Fold the upper layer in half to the right. Fold the lower layer to the left.

3 Fold down a small hem at the top and turn the napkin over.

4 Fold 2 triangles towards the centre, to the height of the hem.

5 Fold the rest of the napkin in 3 and slip it under the 2 collar points.

6 Isn't that elegant!

*Use a large napkin to make this magnificent
decoration. A name card for each of your guests
will slip in easily.*

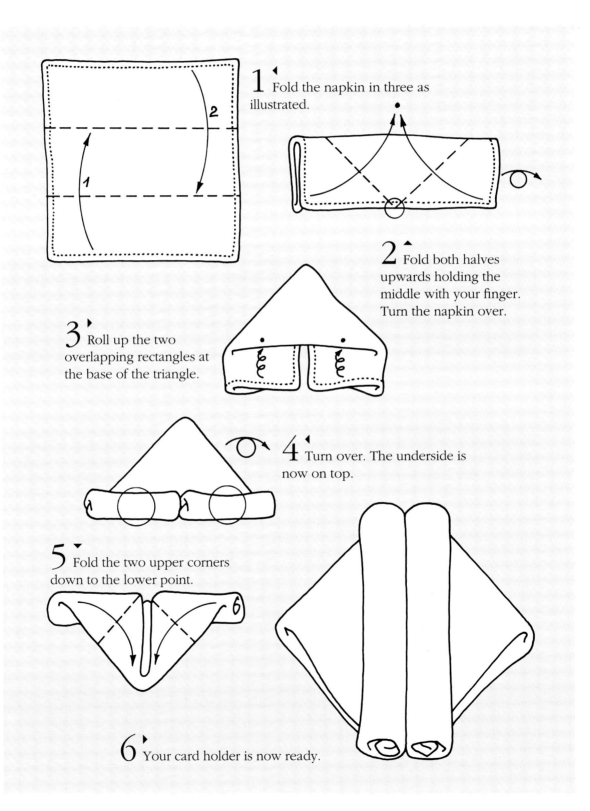

1 Fold the napkin in three as
illustrated.

2 Fold both halves
upwards holding the
middle with your finger.
Turn the napkin over.

3 Roll up the two
overlapping rectangles at
the base of the triangle.

4 Turn over. The underside is
now on top.

5 Fold the two upper corners
down to the lower point.

6 Your card holder is now ready.

*With traditional oriental patience and a
well-starched napkin, amaze your guests with an introduction
to Japanese culture and cuisine.*

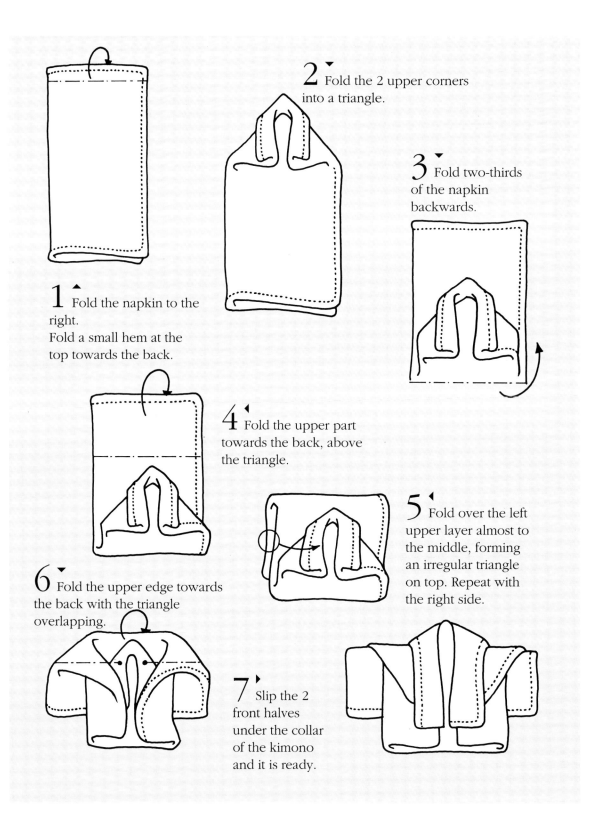

2 Fold the 2 upper corners into a triangle.

3 Fold two-thirds of the napkin backwards.

1 Fold the napkin to the right.
Fold a small hem at the top towards the back.

4 Fold the upper part towards the back, above the triangle.

5 Fold over the left upper layer almost to the middle, forming an irregular triangle on top. Repeat with the right side.

6 Fold the upper edge towards the back with the triangle overlapping.

7 Slip the 2 front halves under the collar of the kimono and it is ready.

*This is a simple but very elegant pattern...
and the Greatcoat will add style
to your finest tableware.*

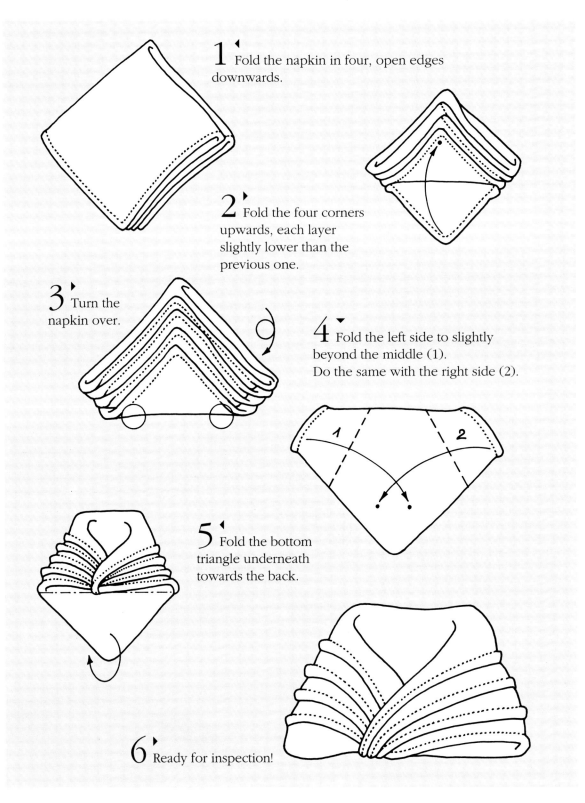

1 Fold the napkin in four, open edges downwards.

2 Fold the four corners upwards, each layer slightly lower than the previous one.

3 Turn the napkin over.

4 Fold the left side to slightly beyond the middle (1). Do the same with the right side (2).

5 Fold the bottom triangle underneath towards the back.

6 Ready for inspection!

UPRIGHT DESIGNS

Would you like your guests to notice your beautifully decorated table as soon as they arrive? Unlike flat designs, where the surprise is kept secret until the beginning of the meal, upright models catch the eye immediately. They prove that you have thought of everything!

Some are purely decorative, such as the fan or the bird. But others are useful too: to hold a name card or to keep a bread roll warm.

Just choose the pattern that goes best with your table plan.

*This mitre will grace your table,
and what is more you can pop
a bread roll or a flower into it.*

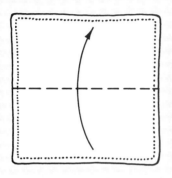

1 Fold the napkin in half upwards.

2 Fold the upper left corner down towards the middle of the base.

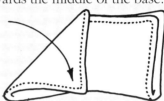

3 Fold the lower right corner up towards the middle of the top.

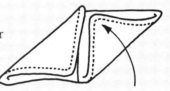

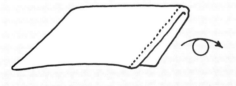

4 Turn the napkin over and place it with the long side towards you.

5 Fold the top layer over and pull out the left point which is underneath.

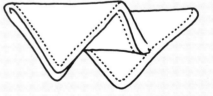

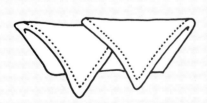

6 Fold the lower right triangle towards the back and then over the left triangle towards the front.

7 Turn the napkin over and slip the two ends one into the other.

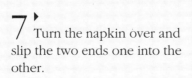

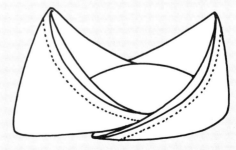

Add real style to your table with this magnificent Spanish fan!
It will be the finishing touch
to your special dinner.

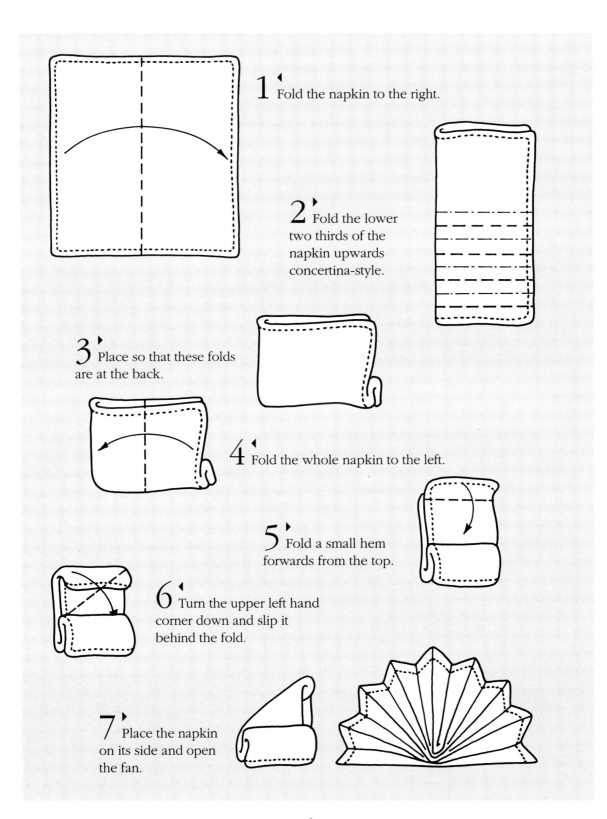

1 Fold the napkin to the right.

2 Fold the lower two thirds of the napkin upwards concertina-style.

3 Place so that these folds are at the back.

4 Fold the whole napkin to the left.

5 Fold a small hem forwards from the top.

6 Turn the upper left hand corner down and slip it behind the fold.

7 Place the napkin on its side and open the fan.

This is a feast for the eyes ...
Your guests will admire it so much that
they won't want to unfold it!

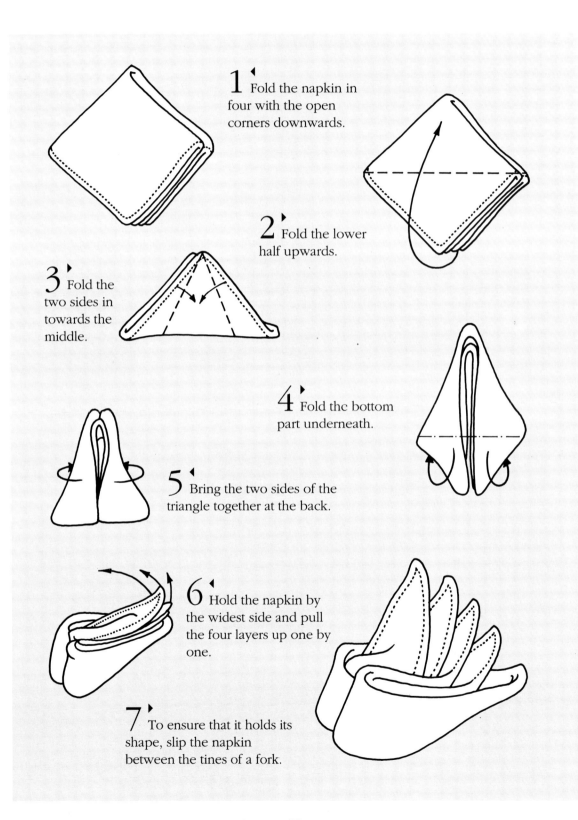

1 Fold the napkin in four with the open corners downwards.

2 Fold the lower half upwards.

3 Fold the two sides in towards the middle.

4 Fold the bottom part underneath.

5 Bring the two sides of the triangle together at the back.

6 Hold the napkin by the widest side and pull the four layers up one by one.

7 To ensure that it holds its shape, slip the napkin between the tines of a fork.

An easy-to-make decoration
using either a cloth or a paper napkin.

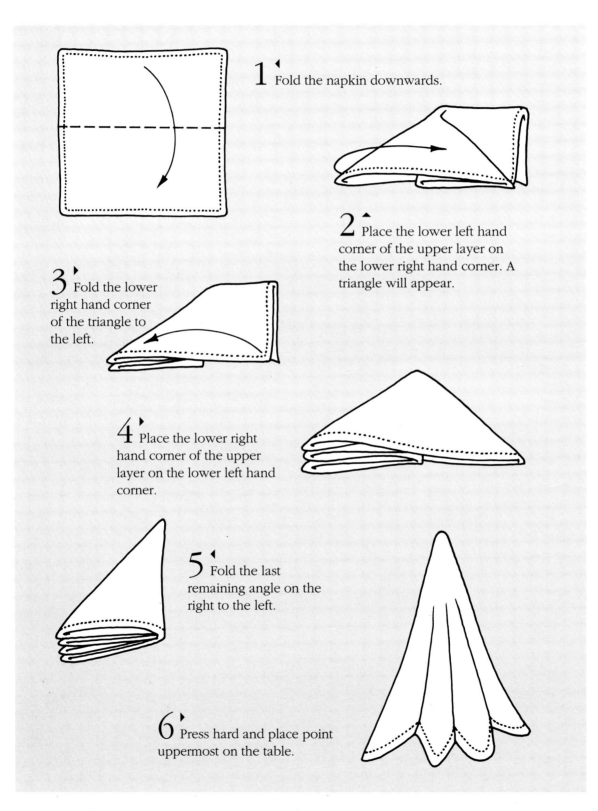

1 Fold the napkin downwards.

2 Place the lower left hand corner of the upper layer on the lower right hand corner. A triangle will appear.

3 Fold the lower right hand corner of the triangle to the left.

4 Place the lower right hand corner of the upper layer on the lower left hand corner.

5 Fold the last remaining angle on the right to the left.

6 Press hard and place point uppermost on the table.

*This fan requires a little more skill
but the result is superb;
a guaranteed success with a delicious meal!*

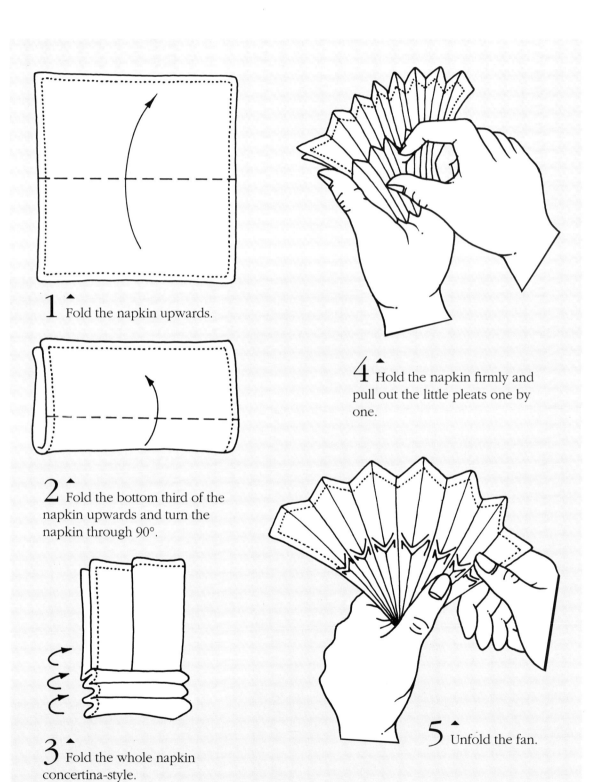

1 Fold the napkin upwards.

2 Fold the bottom third of the napkin upwards and turn the napkin through 90°.

3 Fold the whole napkin concertina-style.

4 Hold the napkin firmly and pull out the little pleats one by one.

5 Unfold the fan.

Why not say it with flowers?
The beauty of the napkin will be
enhanced by the aroma of your meal

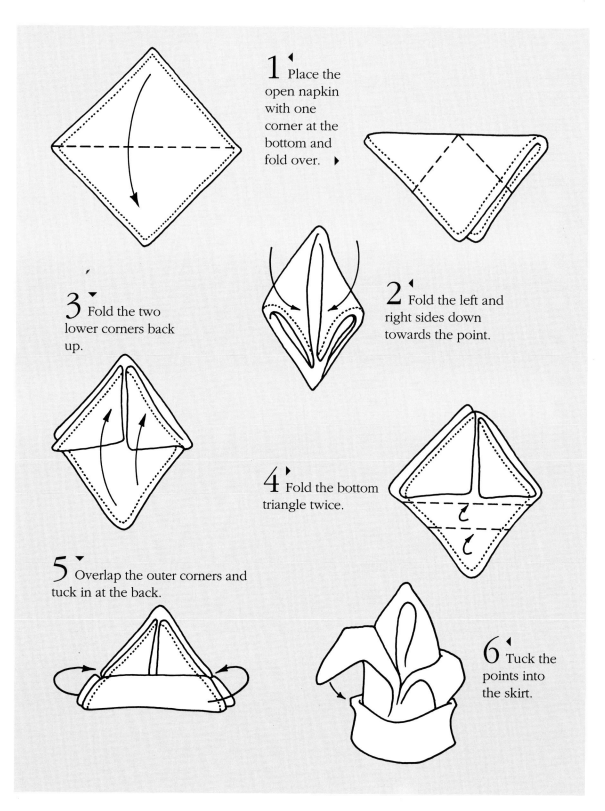

1 Place the open napkin with one corner at the bottom and fold over. ▶

2 Fold the left and right sides down towards the point.

3 Fold the two lower corners back up.

4 Fold the bottom triangle twice.

5 Overlap the outer corners and tuck in at the back.

6 Tuck the points into the skirt.

The Horn of Plenty heralds a fine meal for your guests.
Their mouths will water as soon as they see this napkin
and they will not be disappointed.

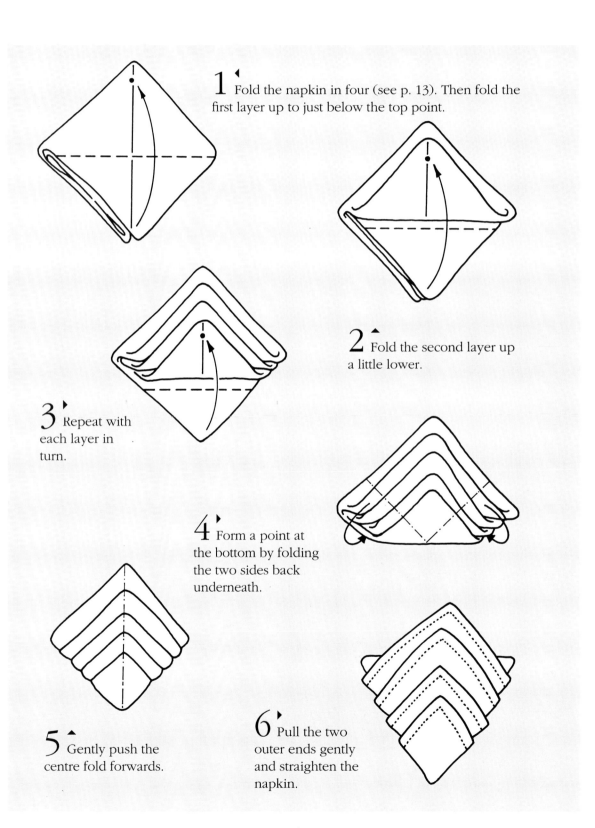

1 Fold the napkin in four (see p. 13). Then fold the first layer up to just below the top point.

2 Fold the second layer up a little lower.

3 Repeat with each layer in turn.

4 Form a point at the bottom by folding the two sides back underneath.

5 Gently push the centre fold forwards.

6 Pull the two outer ends gently and straighten the napkin.

Like a bud opening up on your plate...
A symbol of spring and all the good things in life,
such as a fine meal around a beautifully-laid table.

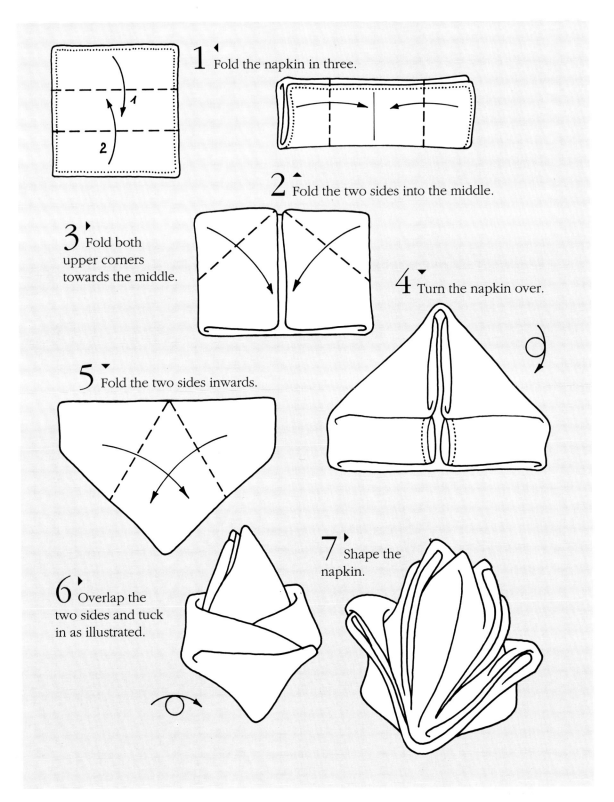

1 Fold the napkin in three.

2 Fold the two sides into the middle.

3 Fold both upper corners towards the middle.

4 Turn the napkin over.

5 Fold the two sides inwards.

6 Overlap the two sides and tuck in as illustrated.

7 Shape the napkin.

49

DESIGNS WITH A NAPKIN RING

Napkin rings are not only practical but decorative and can really enhance your table. Folded designs with a napkin ring are also very easy and quick to make. Ideal for when you have a lot of guests as a simple ring or a bow will add a festive air to your table. No napkin rings at home? Use a glass or a pretty ribbon for maximum effect.

Very simple and very pretty, especially in bright colours!
Instead of a napkin ring you could use
some coloured ribbon to make a little bow.

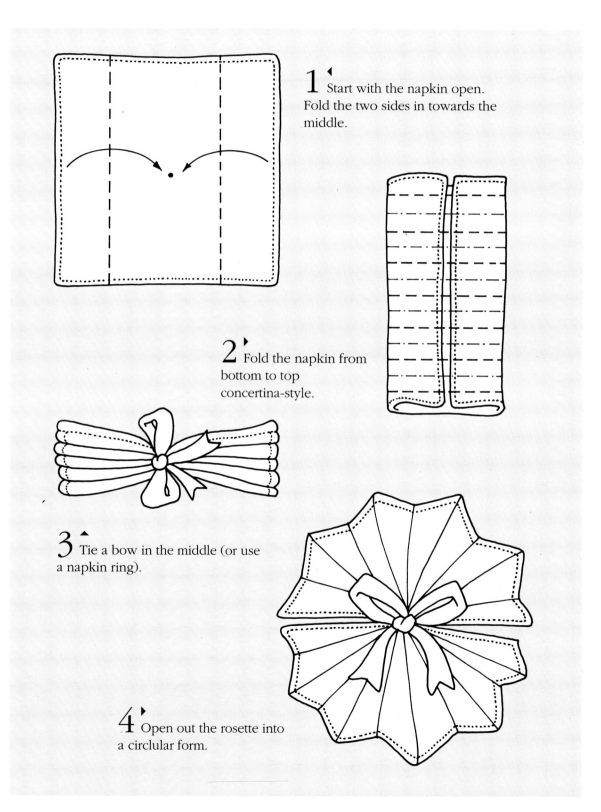

1 Start with the napkin open.
Fold the two sides in towards the
middle.

2 Fold the napkin from
bottom to top
concertina-style.

3 Tie a bow in the middle (or use
a napkin ring).

4 Open out the rosette into
a circlular form.

A bouquet for each of your guests?
Use napkins with a floral motif to give
your table an elegant air of occasion.

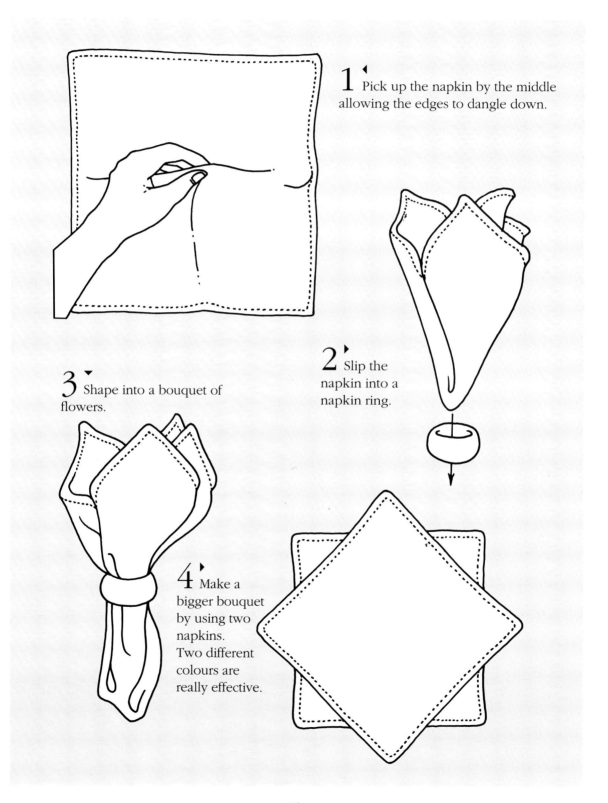

1 Pick up the napkin by the middle allowing the edges to dangle down.

2 Slip the napkin into a napkin ring.

3 Shape into a bouquet of flowers.

4 Make a bigger bouquet by using two napkins. Two different colours are really effective.

As pretty as a picture...
Irises on the table will impress everyone there!
Use your best glasses as vases for these napkins.

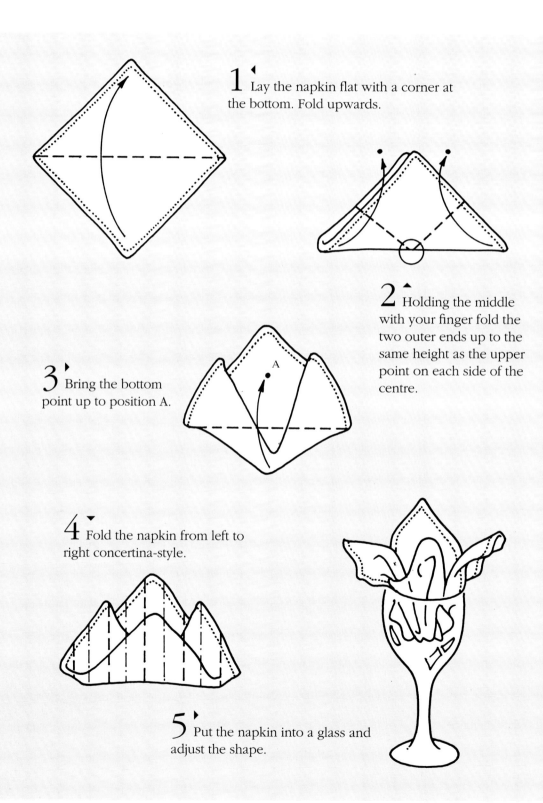

1 Lay the napkin flat with a corner at the bottom. Fold upwards.

2 Holding the middle with your finger fold the two outer ends up to the same height as the upper point on each side of the centre.

3 Bring the bottom point up to position A.

4 Fold the napkin from left to right concertina-style.

5 Put the napkin into a glass and adjust the shape.

*What could be more appropriate for a celebration
than this fan, much-used in Japanese decoration.
It symbolizes well-being and good luck.*

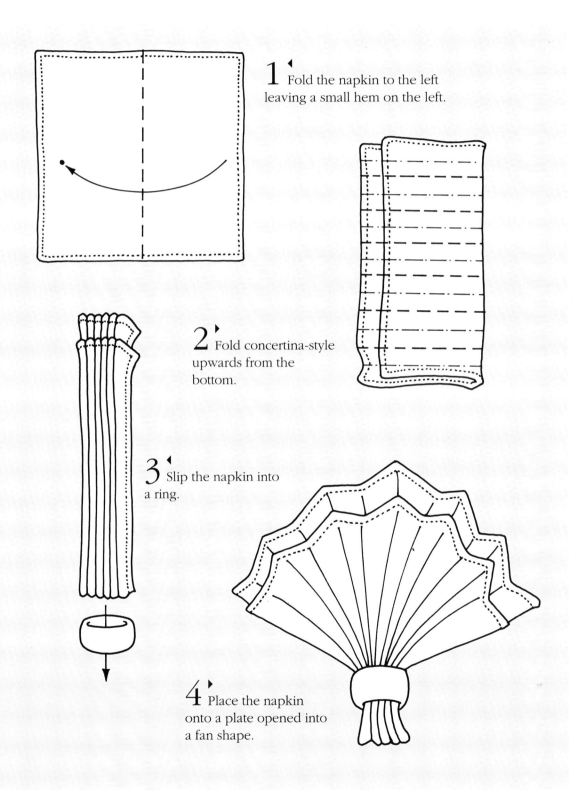

1 Fold the napkin to the left
leaving a small hem on the left.

2 Fold concertina-style
upwards from the
bottom.

3 Slip the napkin into
a ring.

4 Place the napkin
onto a plate opened into
a fan shape.

INDEX OF DESIGNS

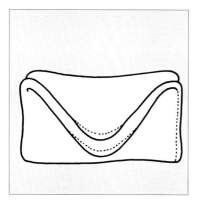

Evening Purse p. 16

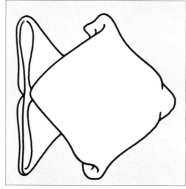

Little Fish p. 18

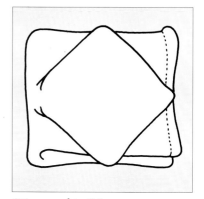

Diamond p. 20

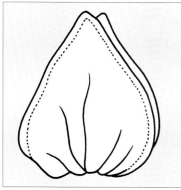

Leaf p. 22

Evening Shirt p. 24

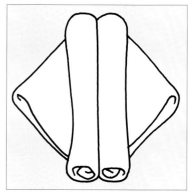

Card Holder p. 26

Kimono p. 28

Army Greatcoat p. 30

Bishop's Mitre p. 34

Standing Fan p. 36

Bird of Paradise p. 38

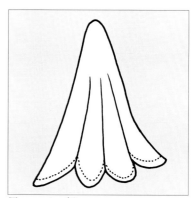

Tower p. 40

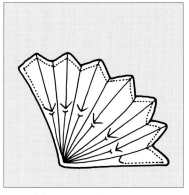

Double Fan p. 42

Fleur de Lys p. 44

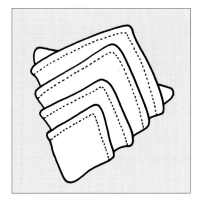

Horn of Plenty p. 46

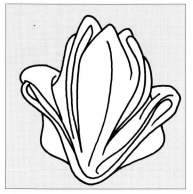

Flower Bud p. 48

Rosette p. 52

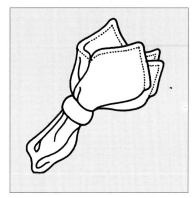

Bouquet of Flowers p. 54

Flowering Iris p. 56

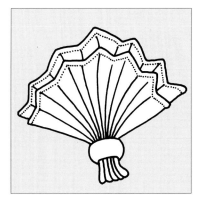

Fan p. 58

This edition published 1994 by Sunburst Books,
Deacon House, 65 Old Church Street, London SW3 5BS

© 1993 by Zuidnederlandse Uitgeverij N.V.,
Aartselaar, Belgium.

© 1994 English translation Sunburst Books

Photography: Steven d'Haens

ISBN 1 85778 133 3

Printed in Belgium